Adama SACKO
Oumar BAH

ANALYSIS OF INTERNAL CONTROL PROCEDURES:

The case of SOMAGEP-SA

ScienciaScripts

Cover image: www.ingimage.com

This book is a translation from the original published under ISBN 978-620-6-72855-9.

Publisher:
Sciencia Scripts
is a trademark of
Dodo Books Indian Ocean Ltd. and OmniScriptum S.R.L publishing group

120 High Road, East Finchley, London, N2 9ED, United Kingdom
Str. Armeneasca 28/1, office 1, Chisinau MD-2012, Republic of Moldova, Europe
Printed at: see last page
ISBN: 978-620-8-34738-3

Adama SACKO
Oumar BAH

ANALYSIS OF INTERNAL CONTROL PROCEDURES:

DEDICATION

I dedicate this dissertation to my dear parents for all the efforts made during my school and student career.

ACKNOWLEDGEMENTS

For the writing of this thesis, I benefited from the contribution and support of several legal entities and individuals, to whom I would like to express my gratitude and sincere thanks, including:

-The management of SUP'MANAGEMENT MALI for its availability

-The teaching staff at SUP'MANAGEMENT in general for the quality of the training I received and in particular my supervisor for his sound advice and unfailing assistance;

-All the staff at SOMAGEP-SA in general and the General Control Department in particular, the Faladiè sales agency and the Supply and Logistics Department, for their availability and the support I received to ensure the smooth running of this work;

-All the members of my family for their unfailing support;

Not forgetting all those people who, in one way or another, contributed to the preparation of this memoir.

SUMMARY

Internal Control plays a vital role in all types of organization, helping to improve the efficiency and economy of certain decision-making processes, minimize and anticipate **risks, ensure the reliability of** economic and financial data, and guarantee compliance and the application of procedures.

In the wake of financial scandals in major multinational corporations, a number of measures have been taken in advanced countries, requiring these companies to set up an Internal Control system and regularly assess its effectiveness against the Internal Control model recognized and recommended by the relevant standards.

The case study focused on the analysis of the Internal Control system of the Société Malienne de gestion de l'Eau Potable "SOMAGEP-SA", in which an attempt was made to analyze the effectiveness of the Internal Control system in place, and to highlight the main risks, by proceeding in stages, starting with a study of existing documentation, take part in internal control missions to carry out control activities, which will enable us to find out about the realities on the ground, interview managers in order to provide answers to the questionnaire on COSO components, and draw up a Work Sheet to assess risks and control operations, analyze answers and findings, and propose recommendations.

The implementation of an Internal Control system, based on well-established concepts and principles, provides companies with a reasonable level of assurance that they are achieving their performance objectives, and that the associated risks are under control, provided that the necessary resources are made available and put into practice.

GENERAL INTRODUCTION

The world of public and private organizations is constantly evolving.
turbulence, risk assessment and uncertainty reduction are major management challenges. In the past, the manager's role was limited to integrating and controlling human resources (the workforce) to keep the company running. Nowadays, however, in view of environmental and socio-economic changes, companies have become so fragile that they arc no longer able to cope with environmental and economic realities.

Faced with this situation, the manager's attention is stimulated, beyond integrating the workforce in the pursuit of maintaining and, above all, developing the organization.

In order to preserve their achievements, companies generally set specific, measurable and time-bound objectives in line with their field of activity.

Any company wishing to achieve its objectives must equip itself with modern management tools adapted to its context.
To achieve these objectives, we need to put in place a series of mechanisms, including "internal control".

Internal control is an essential organizational device, which is set up within the company to ensure that the desired performance is achieved.

Specifically, it aims to ensure compliance and strict application of practices with the legal and regulatory requirements in force, in accordance with the instructions and guidelines issued by the Boards of Directors and General Management.

The proper functioning of a company's internal control process contributes to the safeguarding of its assets, the reliability of its information in general, the control of its activities through the effectiveness of its operations, and the efficient use of its resources.

By helping to prevent and control the risks of failure to achieve the objectives set by the company, the internal control system plays a key role in the management and steering of its various activities.

In view of its importance and the role it plays in a company's operations, it is necessary to periodically assess the effectiveness and relevance of all aspects of the internal control system.

The results will enable the company to correct any deviations in its control system, and to adapt it as effectively as possible, so as to remain competitive and avoid being squeezed out of the market.

Although the Société Malienne de Gestion de l'Eau Potable "SOMAGEP-SA" is a company with no direct competitors and occupying a significant position on the national territory, it must put in place measures that will be adapted to its situation, in order to achieve its strategic objectives, one of which is focused on: "Improving customer satisfaction, while maintaining its reputation and image, in order to increase sales".

In a monopoly situation, achieving this objective requires a great deal of effort and improvement, in particular the mobilization of the financial and technical resources needed to improve the quality of service provided by its staff, whose working conditions must be continually improved to reflect the company's performance.

That's why we've chosen this theme:

"Analysis of the internal control system of the Société Malienne de Gestion de l'Eau Potable (SOMAGEP S.A).

Despite the existence of this internal control system at the Société Malienne de Gestion de l'Eau Potable (SOMAGEP S.A.) since its inception, the system is encountering organizational management difficulties that are affecting its smooth operation.

In an attempt to identify these reasons, the analysis of this theme will focus on the following questions:

1- What are SOMAGEP S.A.'s internal control tools?

2-What is the status of SOMAGEP S.A.'s internal control system?

3-What are the strengths and weaknesses of SOMAGEP S.A.'s internal control system?

The answers to these questions will enable us to analyze the current control system more calmly, and make any necessary suggestions for improvement.

As a result, the notions of internal control will impose themselves on the manager, as they are concerned with the company as a whole.

For this reason, the overall aim of this thesis is to demonstrate the importance of internal control in guaranteeing corporate performance.

Based on this general objective, two specific objectives are defined, namely :

On the one hand, to ensure that the methods and procedures adopted within the company to protect its assets comply with standards, and on the other, to ensure that management instructions are applied.

The study in this thesis will be based on two hypotheses, which will be verified (with reference to the results obtained) and confirmed by the answers obtained from these analyses.

The first assumption is that the internal control system is capable of guaranteeing the company's performance. The second assumption is that the internal control system is capable of adapting to any changes in the environment in order to control risks.

To carry out this work, the methodology will be based on document reviews, interviews and direct observation of departments and managers, in order to identify the company's internal data.

PART ONE: THEORETICAL FRAMEWORK

CHAPTER 1: GENERAL INFORMATION ON INTERNAL CONTROL

This chapter deals with internal control in its entirety, proposing various definitions, and addressing concepts linked to its evolution and objectives.

Section 1: Definition and purpose of internal control

a. Definition

Among the many definitions, two seem to be more complete, namely :

"Internal Control is the set of methods and procedures implemented by the company's management to organize activities, safeguard assets and detect any errors or fraud that may occur in accounting records and financial reporting, and ensure as far as possible compliance with management directives and current laws and regulations, with the aim of improving the company's performance and profitability at all levels".

"A company's internal control is the entire system of controls established by management to conduct the company's business in an orderly fashion, in order to ensure the organization's strategic continuity".

b. Internal control objectives and missions

"Internal Control is a system defined and implemented by the entity under its responsibility. It comprises a set of resources, behaviors, procedures and actions adapted to the specific characteristics of each organization.

In particular, the system is designed to ensure :

- compliance with laws and regulations;
- the application of instructions and guidelines set by General Management or the Executive Board;
- the smooth running of the company's internal processes, particularly those designed to safeguard its assets;
- Reliability of financial information.

Internal Control is therefore not limited to a set of procedures or to accounting and financial processes alone, nor does it cover all initiatives taken by governing bodies or management, such as defining the organization's strategy, setting objectives, making management decisions, dealing with risks or monitoring performance, but provides analyses and recommendations in line with the context and environment.

In a complex and uncertain environment, the company must constantly refocus its objectives and actions. Performance management must be a compromise between adapting to external changes and maintaining organizational coherence to make the best use of resources and skills.

Internal control helps to allocate resources to current strategic priorities through the recommendations made and contributes to optimizing quality, cost and lead times, using all the problem-solving tools available, such as process analysis and analysis of quality management tools.

Internal control helps to manage the social performance variables demanded by stakeholders.

c. Compliance with laws and regulations

These are the laws and regulations to which the company is subject. The laws and regulations in force set standards of behavior that the company incorporates into its compliance objectives.

Given the large number of existing areas of expertise (company law, commercial law, environmental law, labor law, etc.), the company needs to be organized in such a way as to be able to :

- Know the various rules that apply;
- To be informed in a timely manner of any changes made to them (legal watch);
- Transcribe these rules into internal procedures;
- Inform and train employees on the rules that concern them.

d. Application of instructions and guidelines set by General Management or the Executive Board

Instructions and guidelines from General Management or the Executive Board enable employees to understand what is expected of them and the extent of their freedom of action.
These instructions and guidelines must be communicated to the employees concerned, according to the objectives assigned to each of them, in order to provide guidance on how activities should be conducted. These instructions and guidelines must be drawn up in line with the company's objectives and the risks involved.

e. **operation of the company's internal processes, in particular those designed to safeguard its assets**

All operational, industrial, commercial and financial processes are concerned.
The smooth running of processes requires that operating standards or principles have been established, and that performance and profitability indicators have been put in place.
By "assets" we mean not only "tangible assets" but also "intangible assets" such as know-how, image or reputation. These assets can disappear as a result of theft, fraud, unproductivity, errors, poor management decisions or

weaknesses in internal control. Special attention should be paid to the related processes.

The same applies to the processes involved in preparing and processing accounting and financial information. These processes include not only those dealing directly with the production of financial statements, but also the operational processes that generate accounting data.

f. Reliability of financial information

The reliability of financial information can only be achieved through the implementation of internal control procedures, capable of accurately capturing all the operations carried out by the organization.

The quality of this internal control system can be assessed by means of:

- A segregation of duties that clearly distinguishes between recording, operational and conservation tasks;
- A description of the functions used to identify the origins and recipients of the information produced;
- An internal accounting control system to ensure that transactions are carried out in accordance with general and specific instructions, and that they are accounted for in

such a way as to produce financial information that complies with generally accepted accounting principles.

g. Internal control framework

1. The scope of internal control

It is up to each company to set up an Internal Control system adapted to its own situation.

Within a group, the parent company ensures that its subsidiaries have internal control systems in place. These systems should be adapted to their specific characteristics and to the relationship between the parent company and its subsidiaries. In the case of significant shareholdings in which the parent company exercises significant influence, it is up to the parent company to assess the possibility of becoming acquainted with and examining the measures taken by the shareholding concerned with regard to Internal Control.

2. Responsibilities

Management is responsible for raising awareness of the importance of sound management of the company's business and assets.

The division of tasks is of paramount importance for Internal Control: the larger the company, the more responsibilities and authorizations will be assigned to different people, to avoid incompatible influences and functions.

When a new employee takes on a new role, management must ensure that the working methods inherent in this role are communicated to him or her: a good understanding will ensure the efficiency of the process, the reliability of the information and the value of the task.

c. Interpretation

Unlike the accounting system, which captures, records and aggregates transactions to present the results in the financial statements, the Internal Control system comprises methods and procedures that the company adds to the accounting system to achieve a reasonable degree of certainty.

Section 2: Internal control players

- **Internal control players**

All members of staff have a responsibility, to a greater or lesser extent, for Internal Control. It is everyone's business, from the governing bodies to all the company's employees. However, only those who belong to the company are part of the Internal Control system, as they each contribute in their own way to the system's effectiveness.

Third parties may also play a role in achieving the organization's objectives, but simply contributing directly or indirectly to the achievement of objectives does not involve them in the Internal Control system.

- **<u>Board of Directors or Supervisory Board</u>**

The extent to which the Board of Directors or Supervisory Board is involved in Internal Control varies from company to company. It is the responsibility of Executive Management or the Management Board to report to the Board (or its Audit Committee where one exists) on the essential features of the Internal Control system. If necessary, the Board may use its general powers to carry out any further checks and cross-checks it deems appropriate, or take any other initiatives it deems appropriate.

Where it exists, the Audit Committee should carefully monitor the Internal Control system on a regular basis.

To enable it to carry out its responsibilities in full knowledge of the facts, the Audit Committee may interview the head of internal audit, give its opinion on the organization of the department and be kept informed of its work. It should therefore receive internal audit reports or a periodic summary of these reports.

- **General Management / Executive Board**

In any company, the Chairman and CEO bears ultimate responsibility. As such, he is primarily responsible for the Internal Control system. He must ensure that a positive environment exists in which the company's activities and related controls are carried out. He/she must also set an example within the entity.
The same applies to the managers of the company's various functions and units.
General Management or the Executive Board is responsible for defining, promoting and monitoring the internal control system best suited to the company's situation and business. Within this framework, they keep each other regularly informed of any malfunctions, shortcomings or difficulties in application, or even excesses, and ensure that the necessary corrective action is taken.

- **Internal auditors**

Where it exists, the Internal Audit department is responsible for assessing the functioning of the Internal Control system and making recommendations for improvement, within the scope of its missions.
He usually raises awareness and trains managers in Internal Control, but is not directly involved in setting up and implementing the system on a day-to-day basis.

Standards issued by the Institute of Internal Auditors specify that an internal audit must include an examination and assessment of the sufficiency and effectiveness of the organization's internal control system, as well as an evaluation of the tasks assigned to them.

The head of internal audit reports to General Management and, in accordance with the procedures determined by each company, to the corporate bodies, on the main results of the monitoring performed.

- **<u>Company personnel</u>**

To a certain extent, Internal Control is the responsibility of all staff members, and must be mentioned, explicitly or implicitly, in each employee's job description. This responsibility is twofold:

- On the one hand, virtually all employees play a role in carrying out controls. They may have to produce information used in the Internal Control system, or undertake actions necessary to ensure control. The care taken with these actions has a direct impact on the effectiveness of the Internal Control system.
- In addition, all staff members must be required to report to their line manager any problems observed in operations, any breaches of the code of conduct or the organization's internal standards, and any illegal actions. As these actions

may originate from their direct line manager, communication channels other than the usual ones must exist, to enable escalation of such behavior.

It is also vital that all employees concerned have the knowledge and information they need to set up, operate and monitor the Internal Control system, in line with the objectives assigned to them. This is the case for operational managers in direct contact with the Internal Control system, supervisors and financial executives, who must play an important steering and monitoring role.

- **Third parties**

Among third parties, it is generally the external auditors who make the greatest contribution to achieving the organization's objectives in terms of financial reporting and compliance with the various laws and regulations. They provide management and the Board of Directors with an objective, independent viewpoint.

Legislators and supervisory authorities have an influence on Internal Control systems in many companies, either by requiring them to set up control systems, or by carrying out controls directly for some of them.

- **The limits of internal control**

The Internal Control system, however well designed and applied, cannot provide an absolute guarantee that the company's objectives will be met.

The likelihood of achieving these objectives is beyond the company's control. Indeed, there are inherent limits to any Internal Control system. These limits are the result of a number of factors, including environmental factors, uncertainties in the outside world, the exercise of judgment, and malfunctions that may occur due to human failure or simple error.

Furthermore, when setting up control processes, it is necessary to take into account the cost/benefit ratio and not to develop unnecessarily costly internal control systems, even if this means accepting a certain level of risk.

- **Judgment**

The effectiveness of controls is limited by the risk of human error when making decisions that have an impact on the company's operations. The people who make such decisions exercise their judgments in the time available to them, based on the information made available to them, while coping with the pressures of doing business. Such decisions may produce disappointing results, and must be modified in the future.

- **Malfunctions**

When instructions are less well interpreted by staff, their judgment may be faulty, leading to a malfunctioning of the Internal Control system. They may commit errors through inattention or routine. An accounting manager charged with investigating anomalies may fail to do so, or may not pursue his or her investigations sufficiently to take appropriate action, and may be replaced by interim staff lacking the skills required to perform their duties properly. Changes to systems may be introduced before staff have received the necessary training to react correctly to the first signs of a malfunction.

- **Management override of controls**

Since the Internal Control system can only be as effective as the people responsible for its operation, they may override it in order to gain personal advantage, improve the presentation of the company's financial situation or conceal non-compliance with legal obligations. Such improper actions include fictitiously increasing sales, raising the value of the company in anticipation of its sale or a public share issue, underestimating sales or earnings forecasts in order to increase a performance-related bonus... and so on.

That said, breaches of the Internal Control system should not be confused with management interventions aimed at overriding or derogating, for legitimate reasons, from prescribed standards and procedures. In the case of unusual transactions or events, such interventions are generally necessary, and are either carried out openly and documented, or the staff members concerned are notified.

- **Collusion**

Collusion between two or more people can thwart the Internal Control system. Individuals acting collectively to perpetrate and conceal an action may alter financial or management information in a way that cannot be detected by the system.

- **Cost/benefit ratio**

The organization must weigh up the costs and benefits of controls before implementing them.

When assessing the appropriateness of a new control, it is necessary to consider not only the risk of failure and the possible impact on the organization, but also the costs associated with implementing this control.

The costs and benefits of controls are calculated with varying degrees of precision. Generally speaking, it is easier to calculate costs by taking into account the direct and indirect costs of the controls implemented.

Calculating the cost/benefit ratio is further complicated by the correlation between controls and activities. When controls are integrated into management and operational processes, it is difficult to isolate their cost or benefit.

- **The influence of Internal Audit and Management Control on the internal control system**

As early as 1933, Berle and Means pointed out that one of the characteristics of the modern firm was the separation between owners and managers. Managers are in a situation where they must, on the one hand, manage assets belonging to shareholders, and on the other, direct the work of those involved in performance. In large companies, this configuration has given rise to two types of "control":

- External controls requested by the company's owners to ensure that their assets are financially well managed;
- Internal controls set up by management, to ensure that decisions are properly implemented (Management Control) and that control systems are working properly (Internal Audit).

a. Definition of Internal Audit

The IIA has adopted the following definition of internal auditing:

"Internal auditing is an independent and objective activity which provides an organization with assurance about the degree of control over its operations, advises on how to improve them, and helps to create added value. It helps the organization to achieve its objectives by systematically and methodically evaluating its risk management, control and management processes, and by making proposals to improve their effectiveness".

b. Definition of Management Control

"Management Control is the process by which management ensures that resources are obtained and used effectively (in relation to objectives) and efficiently (in relation to the means used) to achieve the organization's objectives".

This decision-making process is one of strategy support and deployment.

c. The role of Internal Audit and Management Control in the Internal Control system

Internal audit and management control are two elements of the internal control system which, although essential, are by no means its entirety.

- **Internal auditing** is an independent, objective activity which uses a systematic, methodical approach to evaluate the control system in place and improve risk management.

The mission of the internal auditor is to assist the company's line managers in the exercise of their responsibilities. To this end, it reports objectively and independently on information, assessments, analyses, opinions and recommendations concerning the activities examined, with a view to improvement. This includes promoting effective control at a reasonable cost, while respecting the principle of proportionality.

- **Management control** is a process that enables a manager to forecast, monitor and analyze the achievements of a program or department, and to take corrective action where necessary. Its aim is to ensure that the resources allocated are relevant to the objectives set, that they are used efficiently in relation to the results achieved, and that the latter are effective in relation to the objectives pursued. In this sense, management control enables performance management (at both operational and financial levels), notably through the use of the following tools:
    - Cost analysis ;
    - Planning techniques and budgeting tools ;
    - Indicators and dashboards ;
    - Comparative analysis.

CHAPTER II: PRESENTATION OF SOMAGEP -SA

This chapter will give a brief presentation of the Société Malienne de Gestion de l'Eau Potable "SOMAGEP S.A" through its history, objectives and missions, as well as its structure.

Section 1: History and

1. History

The problem of access to drinking water for disadvantaged populations is a situation that Mali has been experiencing since its inception. Water is the source of life, and we're always talking about it because not everyone has access to it. That's why successive governments have always put it high on their agendas. Despite undeniable progress, the sector was still evolving in an institutional impasse due to its association with the "electricity" sector.

To correct this cumbersome situation, the State of Mali has initiated reforms in the water and electricity sectors, with a view to improving service. In August 2010, these led to the separation of water and electricity utilities and the creation of two new entities: Société Malienne de Patrimoine de l'Eau Potable (SOMAPEP-SA) and Société Malienne de Gestion de l'Eau Potable (SOMAGEP-SA).

- The Société d'Exploitation de l'Eau potable du Mali, known as Société Malienne de Gestion de l'Eau potable and by abbreviation" SOMAGEP-SA ", is a Société Anonyme d'Etat with a Board of Directors, governed by the laws and regulations in force in the Republic of Mali, in particular the Uniform Act of the Organization for the Harmonization of Business Law in Africa (OHADA) of April 17, 1997 relating to the law on Commercial Companies and Economic Interest Groups, and the relevant provisions of Law N°92 - 002/AN-RM of 27/08/1992 on the Commercial Code.
- The Société Malienne de Gestion de l'Eau Potable was created by Ordinance No. 10-040/P-RM of 05 August 2010 with a capital of 100,000,000 (one hundred million CFA francs). Today, SOMAGEP-SA's capital is 2,000,000,000 (two billion CFA francs), 100% owned by the Malian state.

2. Missions:

The Société Malienne de Gestion de l'Eau Potable, abbreviated to SOMAGEP-SA, is responsible for operating drinking water throughout the leased area:

- Technical, financial and accounting management of the public property required to provide the public drinking water service, based on the criteria defined in the leasing contract with Société Malienne de Patrimoine de l'Eau Potable (SOMAPEP) and the State of Mali;

- Drawing up, planning and carrying out the investments required to extend, rehabilitate and renew the water utility infrastructures delegated to it by the Société de Patrimoine;
- Informing and raising awareness among users of the public water service throughout the leased area;
- And generally, all commercial, industrial, movable property, real estate and financial transactions directly or indirectly related to the above-mentioned purposes and likely to promote their development.

To this end, it is specifically responsible for carrying out the following activities:

a. **Technical, financial and accounting management of public property:**

- Raw water collection and treatment ;
- Pumping and distribution of treated water;
- Water quality control ;
- Reading, billing and customer service ;
- Connection, extension, rehabilitation and renewal of networks;
- Preventive and corrective maintenance of installations;

- Service continuity.

b. Investment development, planning and implementation:

- Programming investments delegated to it by the Société de Patrimoine ;
- Project management and project management for projects and work delegated to it by the Société de Patrimoine.

c. Raising public awareness :

- A communication policy aimed at the general public to encourage the proper use of water, the fight against waste, water conservation, resource preservation and sanitation. This mission can be carried out jointly with the Société de Patrimoine.

3. Organization

Based on the missions listed above, the Société Malienne de Gestion de l'Eau Potable "SOMAGEP S.A" is organized as follows:

A Board of Directors ;

General Management;

A Deputy General Manager ;

Communications Department;

Legal Department;

A Quality Safety Environment Department;

A General Control Department;

Internal Audit Department;

Lab Quality Manager ;

A Budget and Management Control Department;

A Production Department

Electromechanical Maintenance Department

A Distribution Department ;

A Design and Works Department;

A Human Resources Department;

A Finance and Accounting Department;

Sales and Customer Management;

A Purchasing and Logistics Department;

An Information Systems Department.

Section 2: Presentation of the SOMAGEP S.A. Internal Control Department

1. The missions

The missions and activities of the Internal Control Department are as follows:

Carrying out internal investigations and controls to detect concealment and misappropriation of Company assets;

Coordination of the activities of the structures responsible for inspecting the company and for technical control of the work;

Verification of compliance with set management or production principles, norms and standards, for commercial, administrative, accounting, logistics, technical, procurement and legal activities;

Preserve the integrity of the Company's assets and financial resources, and monitor their proper use;

Monitoring, inspecting and reviewing the operation of the Company's activities in order to make findings, identify discrepancies and formulate recommendations;

Supervising the handover of services to the company's operational and functional structures;

Development and implementation of annual programs for the General Control functions (Audit, Inspection & Control);

The initiation of corrective actions or non-conformities with a view to correcting anomalies observed during control operations or inspection missions;

Controlling the processing and closure of open corrective actions and non-conformities, and monitoring the implementation of recommendations to inspected structures.

2. Organization of the Internal Control Department

The internal control department is organized as follows:

- An Internal Control Department Manager
- A works control department
- An anti-fraud service
- An accounting, administrative and commercial inspector

PART TWO: PRACTICAL FRAMEWORK

CHAPTER 3: CASE STUDIES

CASE STUDY 1: Internal control audit at the SOMAGEP-SA sales office in Faladiè.

BACKGROUND

As part of the execution of the Internal Control Department's annual inspection program for 2019, an Accounting, Commercial and Administrative internal control mission took place at the Faladiè Commercial Agency from July 22 to August 05, 2019 following mission order N°424/2019 dated 18/07/2019.

The objective of the mission was to ensure that during the period 01/01/2019 to 22/07/2019, the procedures and work instructions in force were applied in accordance, namely:

Guidelines for accounts receivable clearance

-Compliance of customer payments with recorded revenues

-Regular billing of all agency subscribers

-Efficient use of new meters and correct management of old ones during meter changeover operations

At the end of the mission, recommendations were formulated and implemented.

The various treatments carried out in line with the recommendations made will be presented.

1. Internal control procedures

On the unexpected arrival of the members of the inspection team at SOMAGEP-SA's Faladiè sales office, an emergency meeting was called by the branch manager to introduce the day's guests.

The missionaries presented the mission order and the program of activities, which consisted of proceeding in stages and by sector of activity:

- Accounting activities (cash receipts, cash receipts journals, database entries, regular transmission of accounting documents and archiving of accounting documents);
- Customer activities (process of customer requests and complaints, replacement of old faulty meters);
- Subscriber billing activities (regular billing of subscribers by type of profile (washing areas, standpipes, large consumers, zero consumption);
- Investigation of fraudulent connections.

Each control activity was set up in teams, supervised by a designated controller.

The branch manager then reassured his guests of his complete availability, and called on his colleagues to cooperate fully.

2. PRESENTATION OF THE FALADIE SALES OFFICE AND ITS MISSIONS

a. Definition of a commercial agency

A SOMAGEP-SA Consumer Sales Agency is a local establishment, located within a defined perimeter, to manage all commercial and customer operations, acting in the name and on behalf of the company, by persons employed by the latter.

The various activities carried out within sales agencies are carried out in correlated ways, by people endowed with specific skills for well-defined functions, and empowered by responsibilities that differ according to position in the organization and hierarchical level.

b. Presentation of the SOMAGEP-SA sales office in Faladiè

The FALADIE commercial branch is located on the 30-meter road, 500 meters from the Tour d'Afrique, adjacent to the Marché de Bétail, and has 23,692 subscribers as of April 30, 2021, distributed among the neighborhoods of commune VI of the Bamako district, namely :

-FALADIE

-NIAMAKORO

-SIRAKORO

-DIATOULA

-SENOU

-BANANKABOUGOU BOLLE

-(PART OF SOGONIKO)

-Village CAN

-1008 social housing units

-320 social housing units

The functions that make up a SOMAGEP-SA sales office are :

The Agency Manager :

The Branch Manager is responsible for :

- Contribute to achieving the general objectives of the Sales and Customer Department
- Guarantee SOMAGEP-SA customer satisfaction throughout the agency's perimeter.

The Branch Manager is the first line manager and superior of all branch staff, coordinating and supervising all branch activities (customer relations, billing and collection), and representing the company in relations with third parties within the branch perimeter.

He has good technical and managerial skills, with exemplary aptitudes.

- **Billing Division**

Responsible for the administrative management of the agency's sales, it handles the billing of subscribers' consumption and deals with all complaints arising from them, and is made up as follows:

-Billing manager

-Billers (3)

-Zone agent (13)

- **Customer Management Division**:

Responsible for customer management, it handles the collection of consumption invoices and invoices for reimbursable work (estimates), the recovery of unpaid invoices, and the receipt and processing of various customer requests and complaints. It is composed as follows

-Customer Manager

-Collection agent (3)

-Greeter (3)

-Cashier (3)

-Branch Manager (1)

-A driver (1)

-A Technical Sales Agent (1)

-An accountant (1)

3. PRESENTATION OF FINDINGS, RECOMMENDATIONS AND TREATMENTS

In Table 1 below, this section presents the anomaly findings and recommendations made, and in the second table (2), the action taken on the various recommendations and the people responsible for dealing with them.

TABLE 1: ANOMALY FINDINGS

This table will present the various anomalies found during the course of the inspection and will indicate who is responsible for dealing with the anomalies found.

ANOMALY REPORTS FOR THE BRANCH MANAGER

N°	Description of anomalies	Recommendations
01	Total lack of safety in the storage of new meters, and also of meters that have been changed while awaiting disposal.	In the absence of a suitable room, find a secure cabinet to store all new and old meters under the responsibility of the RC.

FAULT REPORTS FOR THE CUSTOMER MANAGER

(Non-exhaustive)

N°	Description of anomalies	Recommendations
01	Presence on the change file (project) of 15 new meters that are not on the list of meters provided by the Metering Division.	Elucidate the origin of these meters and justify their use.
02	Failure to return to the Metering Division 25 new meters not used during large-scale changeover operations.	Justify the actual use of these meters or, failing

		that, return them to the Metering Division.
03	Inadequate filling of cash books. The physical count of the day's takings is written instead of the previous balance.	Fill in cash register books correctly, without erasures or correction, taking care to write each item of information in the right place.
04	The presence of erasures, overwriting or even the use of correction fluid (blanco) on notebook pages.	
05	Non-participation of the RC in the cash desk closing; he signs the notebooks only after the closing.	Participate in cash closures in accordance with DRC IT 01-01 Invoice collection and cash cancellations.
06	Non-stop cashier : • Box I: from 02 to 03/01/2019 ; • Box II: 28/02 and 02/05/2019 ; • Caisse III: 15/01 and 14/02/2019.	Daily closing of cash registers in accordance with DRC IT 01-01 Invoice receipts and cancellations.

ANOMALY REPORTS FOR THE BILLING MANAGER

N°	Description of anomalies	Recommendations

01	04 customers billed for closed doors.	Send customers a letter of discharge, in which you explain the difficulties of succession, your expectations and the steps you will be obliged to take if the situation persists.
02	13 customers billed at a flat rate for inaccessible meters (underground).	Write to the DD to ask for underground meters to be raised.
03	28 flat-rate customers for blocked meters.	Confirm the blockage and replace the meters if the cases are confirmed.
04	09 flat-rate customers billed for illegible meters.	Replacing meters
05	03 customers billed at a flat rate for faulty meters.	
06	03 customers billed for maintenance on uninhabited homes.	Educate customers not to accumulate maintenance bills without paying them.

TABLE 2: MISSION MEMO

N°	Job description	Manager
01	**METER CHANGE COMPONENT** • Confirm or correct the old meter numbers marked in red in the CTR N° column and do the same for the site or customer references (see René's e-mail sent to the RC on 29/07/2019); • Update the 2019 meter change file initiated by the agency, as eleven (11) old meters counted in the ATC office do not appear in the file (cf. e-mail from René sent to the RC on 29/07/2019) ; • Collect and scan the scrapping sheets for meters changed solely by the Faladié branch in 2019, regardless of the changeover project carried out by the companies; • Send all the above information and documents to the inspection department by e-mail.	Branch Manager, Customer Manager, Billing Manager
02	**CUSTOMER SECTION** • Provide a summary table of the Agency's unpaid invoices using the file provided by the inspection department;	

	• Provide an updated detailed statement of unpaid invoices for hydrants (BF) and carwashes (AL) from January 1er to July 31 2019; • Provide an update on the Agency's 20 most indebted subscribers; • Provide a summary table of cancelled cash receipts from January 1er to July 31 2019, specifying for each case the reason for cancellation and the cashier who requested it; • Provide a summary table of subscription cancellations and transfers, specifying the balance of the departing customer.	

TABLE 3: RECAP TREATMENT OF RECOMMENDATIONS

The table below sets out the recommendations made in connection with the anomalies identified during the audit, and the action or proposed action taken on these recommendations.

These recommendations were followed up over the course of a year. An additional mission will be scheduled to assess the implementation of these recommendations.

Recommendations	Implementation Manager	Treatment or proposed treatment
1) Find a suitable room or, failing that, a secure cabinet, to store all new and old meters under the responsibility of the RC.	Branch Manager	1. Provision of a secure metal cabinet in the ATC office (see photo of cabinet).
2) Compulsory participation in cash desk closures in accordance with Work Instruction 'IT DRC 01-01 Collection of invoices and cancellation of collections;	Customer Manager	2. Effective participation of the Customer Manager in each cash desk closing (cf. one page per cash desk per month)
3) Validate booklets immediately when checkouts stop;		3. All checkout books are systematically validated by the Customer Manager and cashiers at checkout.
4) Ensure that cash register books are filled in correctly, without erasures or "Blanco";		4. All errors and overwritten entries are written out in full on a new page of the cash book.

5) Comply with Work Instruction DCC-DRC IT 01-01 Collection of consumption and work invoices and collection cancellations by printing confirmation messages and the originals of cancelled receipts for archiving purposes;		5. A chrono is opened for archiving supporting documents for cancellations.
6) Create a summary table to track collection cancellations ;		6. The cancellation tracking table is updated and checked weekly by the Agency Manager (see cancellation tracking table).
7) Moratoriums must be drawn up carefully and precisely to avoid any ambiguity;		7. Moratoriums are drawn up in accordance with the information to be entered in the current work instruction (see moratorium copy).

8) Conclude moratoria to settle unpaid debts;		8. Moratoriums are systematically concluded in accordance with current work instructions (see moratorium copy).
9) Write the full identity of each signatory on the moratoria;		9. Moratoriums are drawn up in accordance with the information to be entered in the current work instruction (see moratorium copy).
10) Precede the customer's signature with the words read and approved ;		10. The words "**read and approved**" will appear on future moratoria (cf. no moratoria have yet been concluded).
11) Draw up a moratorium amortization table containing the following information: the amount of the debt, the amount of each installment, the various due dates, the total paid and the balance after each payment;		11. the amount of the debt, the amount of each instalment, the various due dates, the cumulative amount paid and the balance after each payment are indicated on the moratoria in accordance with the current work instructions (see moratoria copy).

12) Check and ensure correct and regular updating of the meter change file		12. The meter change table is updated daily and checked by the RC (see meter change monitoring table, list of meters to be changed and Service Order).
13) Elucidate the origin of the **15 new meters** not on the list of meters made available to the agency by the Metering Division;		13. All meters made available to the agency are entered in the meter change tracking table (see meter change tracking table).
14) Complete point 1 of the memo on changing the meter;		14. Point 1 of the memo has already been dealt with.
15) Justify the effective use of the nine new meters absent from the agency and the meter change file;		15. (see meter change follow-up table, copy of BIs and screenshots of meter records in the eGEE database)
16) Find and scrap the 137 meters changed but physically absent from the branch, plus		16. All rebus meters have been sent to the meter department (see download copy of the

the 11 changed meters inventoried at the branch.		list of change meters sent to the meter department).
17) Correctly check books at checkout, before validation;		17. Till books are checked during check-out stops before any validation (cf. copy of till book for the last 5 days of check-outs).
18) Daily compliance with Work Instruction DFC01-01 Cash closure, collection and disbursement of funds.	Accountant	18. the accountant ensures daily compliance with work instruction DFC01-0 Cash closure, collection and disbursement of funds.
19) Send customers whose doors are always closed a letter with a discharge, in which you explain the difficulties of resuming work, your expectations and the measures you will be obliged to take if the situation persists;	Billing Manager	19. Correspondence is regularly sent, but often without discharge, as the occupants are not on site. (cf. copies of correspondence)

20) Request DD in writing to raise underground meters;		20. The list of meters to be upgraded is forwarded to ATC for work (see email, list of meters).
21) Confirm blocked meters and replace proven cases;		21. Meters with anomalies are confirmed by ATC before replacement (see table for monitoring meter and BI changes).
22) Replace unreadable and defective meters.		22. The list of faulty meters is forwarded to ATC for replacement by the RA and RC, who schedule the activity (see meter replacement tracking table).

Analysis: It was found that customer and sales management procedures are respected. On the other hand, it has been noted that procedures are not applied over time, due to a lack of rigor in monitoring activities.

CASE STUDY 2

1. BACKGROUND

Pursuant to mission order N° 457/2020 dated 05/10/2020, the General Control Department carried out an inspection mission at the Purchasing and Stocks Department (DAS).

This mission is part of the implementation of the inspectorate's 2020 annual plan. The selected period runs from January 1, 2019 to November 23, 2020.

2. OBJECTIVES :

The aim of this mission is to ensure :

- That all purchases of goods and services are made in accordance with PO 015-02 PURCHASES AND SUPPLY;
- That stock management is carried out in compliance with PO 014 STOCK MANAGEMENT ;
- That the store procurement process helps control costs and avoid stock-outs;
- Theoretical inventory quantities are identical to physical inventory quantities;

- All inventory transactions have been regularly and correctly recorded in SIGA;
- That storage conditions guarantee the security of stock against deterioration and theft.

3. **METHODOLOGY** :

It consists of an on-site, independent examination of purchasing, procurement and inventory management operations, from purchase requisitions, purchase orders and contracts. The same applies to inventory management, for goods in, goods out and transfers/shipping.

This approach ensures that movements are traceable, and that there is no collusion, embezzlement, over-invoicing or overstocking.

Given the diversity and size of the inventories, and the large number of purchasing and supply operations, the work was carried out in the following stages:

- Sample audit of deliveries of goods and services in 2019 and 2020, based on orders and contracts ;
- Sampling verification of physical and theoretical inventory balances ;

- Verification of stock movements based on the various documents initiated for this purpose (supplier delivery note, supplier order receipt, returns note, goods issue note, transfer note and dispatch note);
- Checking the cost of stock transfers between warehouses ;
- Verification of storage conditions for certain goods in relation to their security.

4. PRESENTATION OF THE PROCUREMENT AND LOGISTICS DEPARTMENT

4.1. MISSIONS

Reporting to the General Manager, the Procurement and Logistics Department will be responsible for managing and coordinating procurement and logistics activities. The department's new strategic orientations include

- Support SOMAGEP-SA's various entities in terms of procurement and logistics;
- Drawing up and updating SOMAGEP-SA's equipment purchasing and renewal policy;
- The definition of a cost control policy for supplies and services;

- The definition of a policy to control costs and deadlines for the acquisition of equipment and services;
- Optimizing procurement of supplies, services and inventory management in terms of quality and rationality;
- Inventory management and item referencing ;
- Keeping inventories of connection and network equipment, electromechanical parts, chemicals, spare parts, office supplies, consumables, etc;
- Development and implementation of a fleet maintenance policy;
- Optimizing and monitoring reform and asset disposal processes;
- Property management.

4.2. ORGANIZATION

To carry out its mission, the Supply and Logistics Department will now be supported by the following Departments:

- **Procurement Department**: Its main mission is to conduct and conclude all purchasing processes for any product or service required by SOMAGEP-SA for the production, transport and distribution of drinking water, under the best possible conditions of quality, price and delivery times, while ensuring greater profitability for the business.

- **Stock Management Department**: Its main task is to ensure that stocked items (chemicals, connection materials, network materials, spare parts, etc.) are available at all times, so as to avoid any stock shortages that could jeopardize service continuity.
- **Logistics and Asset Management Department**: Its main mission is to manage everything to do with transport, the vehicles needed for transport, warehouse management, handling, etc., by optimizing their circulation, monitoring and maintenance of general resources (rolling stock, fuel, administrative and on-call buildings, assigned telephone resources, etc.) at the lowest cost and within the shortest deadlines, and with maximum safety.

5. FINDINGS

Given the sensitive nature of the information arising from the audit findings, financial information will not be disclosed. On the other hand, administrative information will be mentioned in the presentation and treatment of the recommendations.

6. PROCESSING RECOMMENDATIONS

Proposed processing of Inspection Recommendations CCA 2020-DAS

N°	Recommendations	Treatment or proposed treatment	**TREATMENT LEVEL**
1.	Scrupulously comply with PO 015-02 Purchasing and Supply, so that the principle of accrual is the exception rather than the rule.	- The procedures for soliciting suppliers will be scrupulously respected, involving all those concerned. - Application of PO 015-03 to replace PO 015-02	**Effective- Application PO 015-03 provide copy of PO 015-03**
3.	Please ensure that the bid analysis report is included in the files.	The analysis reports will be included in the files in accordance with the OP.	**Effective provide order list 2021**
4.	It is imperative that the bid analysis report be validated by all participants.	The analysis reports are validated by all participants.	**Effective supply of acceptance certificates 2021**
5.	Please validate the quality of the delivered material with the user.	Receipt slips are systematically validated for quality by users	**Effective**

6.	Meeting supplier delivery deadlines	Delivery times are adhered to and contractual provisions are applied	**Effective.** **At the DAL, one agent is currently responsible for monitoring supplier deadlines.**
7.	Replace the lock on the outside door of the bursar's office, or even reinforce the device to make the premises more secure.	The lock has been replaced	**Effective**
10.	Approve BTs in accordance with PO 014-02, or delegate this authority by memorandum to another manager with direct authority over the assigning organization. Take into account the contents of the memo in the next revision of PO 014.	PO 014-02 is strictly applied.	**Apply PO 014-02 Inventory Management in its pt. 4.3**
11.	Install a grid door between the warehouse and the product	The separation is made	**Effective**

	preparation room to secure stored products.		
12.	Repair holes in the roofs of salt and hydrated lime stores to stop rainwater leaks.	Repaired roof	**Effective for the chemicals store and work-in-progress for the djicoroni para calcium hypochloruria shed**
15.	Define a minimum safety stock for items as part of inventory management.	Existence of minimum security stock for items	**Effective**
17.	Correct all negative balances in SIGA.	Correction in progress	**Permanent treatment**
19.	Update entry of all stock receipts in the warehouse.	Updating entries	**Permanent treatment**
24.	Update all stock records on an ongoing basis	Stock cards are up to date	**Permanent treatment**

25.	Create stock cards for items in the warehouse that don't have any.	All items in the store have a stock card	**Permanent treatment**
28.	Make entries on the stock sheet in accordance with the order.	Stock entries are made in accordance with PO 014	**Outstanding**
29.	Write on the BS, without erasure or overwriting, the quantities served and those not delivered.	Discharge slips (DS) are issued in accordance with PO 014	**Effective**
30.	Reject any exit slip on which the description of all or part of the items is incomplete.	All BS with incomplete descriptions are systematically rejected.	**Effective**
31	Calculate the undelivered quantity correctly, so that it is always the difference between the quantity requested and the quantity delivered.	Unserved quantities are correctly calculated.	**Effective**
32.	Always serve the requested quantity of A4 copy paper,	A4 paper is always measured in reams.	**Not included**

	systematically considering the ream as the only reference unit.		
34.	Serve what has been requested and transfer what has been served in such a way as to ensure consistency between the information contained on the documents used.	Strict application of PO 014	**Effective**
35.	Send the white, yellow, green and blue sheets of the BT bundle to receiving warehousemen, in all item transfer operations.	Application of IT Gestion Bon de Sortie.	**Effective**
36.	Ask the receiving warehousemen to keep the yellow sheet of the BT and to return the other sheets of the bundle (white, green and blue) to the issuing warehouseman; duly completed with the quantities received, dated and signed;	Application of IT Gestion Bon de Sortie.	**Effective**
38.	Reject any exit slip on which the reason for exit is not specified.	Application of IT Gestion Bon de Sortie.	**Application Service note**

			n°2021/30/DG/DAL of 23/11/2021 Validation of BS
43.	Always write the date of transfer and the identity of the carrier on each BTr.	Application of IT Gestion Bon de Sortie.	**Effective**
44.	BTs must be approved by the DAS, unless a memo authorizing another manager to do so is issued.	Application of IT Gestion Bon de Sortie.	**Application PO 014 02**
48.	Ask receiving warehousemen to return the original BT, duly completed with the quantities received, dated and signed, to the issuing Chemicals warehouseman.	Application of IT Gestion Bon de Sortie.	**Effective. Permanent**
49.	Correctly fill in stock cards with all the information they should normally contain.	Application of PO 014	**Effective**
51.	Record all goods-in and goods-out operations in real time on	Application of PO 014.	**Apply PO 014 02**

	stock cards and in SIGA, in compliance with PO 014.		
53	Group similar items together in one place in the store.	In progress	

Analysis: at this level, the assessment shows that purchasing procedures are well respected, and that there are no shortcomings in monitoring the consumption of equipment made available to users.

However, it has been noted that procedures for entering stock sheets and archiving them are poorly applied. There is also insufficient information on the identification of users and the nature of the work to be carried out by these same users, for statistical purposes on the most consumable materials, and the absence of a manual and method for storing materials.

CHAPTER 4: ANALYSIS OF THE INTERNAL CONTROL SYSTEM OF THE SOCIETE MALIENNE DE GESTION DE L'EAU POTABLE "SOMAGEP S.A" (MALIAN DRINKING WATER MANAGEMENT COMPANY)

In this chapter, the stages in the internal control assessment process will be described, and conclusions drawn on the analysis of internal control at SOMAGEP S.A. will be given.

Section 1: Analysis of the internal control system of Société Malienne de Gestion de l'Eau Potable "SOMAGEP S.A".

SOMAGEP-SA's internal system is based on procedures and work instructions established to provide a framework for its operations.

This involves analyzing the description of these systems and procedures, the confirmation of the system's understanding, the preliminary assessment of internal control, the confirmation of the application of the system's strengths, the final assessment of internal control, and the internal control assessment report.

1. Assessment of the existence of internal control at Société Malienne de Gestion de l'Eau Potable "SOMAGEP S.A".

The auditor acquaints himself with the company's internal control system, endeavoring to grasp all the methods and procedures relating to its organization. To this end, he uses a descriptive memorandum or flow charts to assess the existence and effective operation of internal control.

2. Familiarization with the internal control system

This stage consists of checking that the control system is functioning correctly, by verifying the descriptions received. Compliance tests (or comprehension tests) are used for this purpose.

3. Evaluating the existence of internal control

This involves carrying out an initial analysis of internal control, normally based on a questionnaire and analysis grid.

The questionnaire enables the auditor to make as complete an observation as possible on each of the points submitted for critical judgment. To achieve this, the QCI must include questions that are relevant to the complete observation.

a. Methods and choices

Theoretical internal control strengths and weaknesses will be determined at this stage. It is planned to carry out a diligent analysis of risks linked to the control environment, and also to analyze specific risks by cycle, with the aim of highlighting weaknesses in procedures in order to target controls.

b. Objectives

The purpose of the internal control questionnaire is to evaluate the internal control system applied by the organization. It highlights, by cycle, the questions which enable us to assess the degree of control over the procedures applied.

The purpose of this questionnaire is to assess the organization's existing internal control system, with a view to deducing its strengths and weaknesses, and then to make suggestions regarding internal control weaknesses.

c. Targets and reasons for choosing them

The questionnaire is submitted to the people who hold positions of responsibility within the Société Malienne de Gestion de l'Eau Potable "SOMAGEP S.A", by signature or, decision, or authorization, during the interviews which will be held with the persons in charge.

What these targets have in common is that they exercise a degree of control over the implementation of the procedure adopted by the organization.

The choice focuses on control points, as it is felt that there is more risk if these control points are not properly managed by those in charge.

2. Questionnaire

The Internal Control Questionnaire is a tool for carrying out the internal control assessment phase. It is an analytical grid designed to enable the auditor to assess the level of internal control within the audited entity or function, and to provide a diagnosis.

The methodology adopted in developing the questionnaire was based on theoretical notions of understanding internal control, to enable the system to be assessed. It has been designed so that various questions can be answered with "YES" or "NO". YES" answers correspond to strong points, indicating that the company theoretically has the appropriate measures in place to achieve its internal control objectives. The "NO" answers correspond to weak points and concern procedural shortcomings. A column dedicated to comments is placed last, to allow in-depth analysis of answers that are more open-ended than others.

Table 1 <u>Internal control questionnaire</u>

QUESTIONS	ANSWERS		OBSERVATIONS
	YES	NO	
1. Are ethical and integrity values formally	YES		A code of ethics and professional conduct has been drawn up and implemented.

defined within SOMAGEP?			
2. Are the staff performing the activities competent?	Yes		Personnel are selected on the basis of the criteria defined in the job descriptions, in accordance with the needs expressed.
3. Do the values of ethics and integrity match the company's management styles?	Yes		The values enshrined in the code of ethics and deontologies are congruent with the company's management style, having been drawn up by the staff themselves.
4. Are responsibilitie s defined?	Yes		Responsibilities are clearly defined on an organization chart available in each department of the company, in accordance with the company's skills repository and on job descriptions.
5. Do agents have job descriptions?	Ye s		Every member of staff, at whatever level, has a job description defining his or her tasks, responsibilities within the process, and the skills required to perform his or her duties.
6. Is there a company	Yes		The company's general organization chart

organization chart?			defines its missions and functions.
7. Is this chart regularly updated?	Ye s		The company's organization chart is regularly updated in line with new developments.
8. Is there a risk map? If so, are they updated?	Yes		Risk mapping exists and is updated regularly
9. Is the company's internal control system assessed?		no	The system has not been evaluated since the company was set up.
10. Is internal control a regulatory requirement?	yes		To ensure the smooth running of the company, an objective, an action plan and a program of activities are validated each year by the company's general management.
11. Is there an internal procedures manual?	Yes		Each process has its own procedure manual for the activities it carries out.
12. If yes, is it regularly updated?	Yes		It has been updated in line with developments and recommendations arising from various

			control and audit missions.
13. Does your organization have its own audit charter?	Yes		The company has an audit charter that is updated each time a new development takes place.
14. Does your organization have an audit department?	Yes		There is an Internal Audit department
15. is an annual internal control report prepared?	Yes		An annual report on the control activities carried out is provided after each financial year.
16. Are these control measures constantly adapted?		No	Control measures are inadequate due to lack of resources
17. Are payment vouchers signed on the basis of	Yes		The documents are checked in accordance with the procedure in

supporting documents?			force for validating expenditure.
18. Is cash on hand maintained at a minimum level?	Yes		A minimum cash float of 50,000 FCFA is available at each caisse.
19. Is there a regular check of existing checkouts?	Yes		At the end of each collection day, the cash register is checked by means of a cash stop between the accountant, line managers and the cashier, in accordance with a work instruction governing the activity in question.
20. Are cash journals up to date?	Yes		An updated statement of the company's accounts is drawn up at a given frequency, to ascertain the company's financial flows, in accordance with a work instruction in force.

21. Are they regularly reviewed by a manager?	Yes		Reconciliations are regularly reviewed and validated by the Finance and Accounting Manager.
22. Does a competent and clearly defined manager regularly review reconciliation statements?	yes		Reconciliations between journals and bank statements are performed by the managers of the various participants in the process.
23. Are cash flow forecasts regularly monitored?	yes		Cash flow forecasts are monitored by the Budget and Management Control Department.

3. Results

The results obtained from the collected responses are compiled and analyzed more calmly.

To this end, the strengths and weaknesses of the internal control system deduced from the above analysis of the internal control questionnaire will be presented.

In the light of the results obtained, it can easily be said that the internal control system of the Société Malienne de Gestion de l'Eau Potable (SOMAGEP-SA) is effective. It relies on tools, notably procedures and work instructions, enabling it to manage activities and minimize risks.

Section II: Internal control analysis report

In this section, the final assessment of SOMAGEP-SA's internal control system is presented, followed by the conclusion of the internal control analysis.

1. Final analysis of internal control

At this stage, the auditor is in a position to distinguish between strengths and weaknesses; together, these elements provide the basis for his definitive assessment of internal control, which he sets out in a summary document (or system assessment table).

This table summarizes the general points to be checked on the structure's internal control in order to be able to issue the internal control assessment report, highlighting the questions which enable internal control to be assessed.

a. Methods and selection

The definitive internal control analysis table is used to highlight the most important issues to be remembered in the analysis that has been carried out previously. Important points need to be recapitulated in order to be able to issue the conclusion on the internal control analysis.

The table is also presented in such a way as to highlight the degree of risk incurred in the event of internal control weakness in the area concerned.

b. Objectives

The aim of this definitive analysis is to set out the level of risk in relation to the points concerned. It also aims to summarize the key points of the internal control questionnaire as a whole.

c. Targets and reasons for choosing them

The targets are the same as in the internal control questionnaire. Positions of responsibility have been chosen because they are control points that need to be assessed.

2. Analysis of results

The results are processed in the same way as in the internal control questionnaire. YES" answers correspond to strong points and indicate that the company theoretically has the appropriate measures in place to achieve its internal control objectives, while "NO" answers correspond to weak points and concern procedural shortcomings.

The word "risk" has been added to the table to show the degree of risk incurred by the company in the event of internal control weaknesses.

a. Final analysis of internal control

GENERAL INTERNAL CONTROL ENVIRONMENT	Yes/ No	Risks: (low; medium; high)	Comments
Has the internal control assessment revealed the existence of numerous internal control weaknesses?	Yes	Average	The strengths and weaknesses of internal control were highlighted by the results of the questionnaire.
Is the organization plan inadequate, as evidenced by the absence of an organization chart and procedure manual?	no	Low	The procedure manual and organization charts have been identified.

Are there any situations or events that suggest the existence of fraud or error leading to material misstatement?	No	Low	Despite some shortcomings in the day-to-day management of activities, it has been noted that agents regularly follow up on their
Is management aware of the need for effective internal control?	Yes	Low	This is demonstrated by the existence of the Internal Control department, which carries out its activities on a full time basis in

b. Results

The results of this final assessment are presented below. As explained in the processing of the results, "YES" answers correspond to strong points and indicate that the company theoretically has the appropriate measures in place to achieve its internal control objectives, while "NO" answers correspond to weak points and concern procedural shortcomings.

The Internal Control system is well designed and well applied, but however available it may be, it cannot provide an absolute guarantee that the company's objectives will be achieved. The probability of achieving these objectives does not depend solely on the company's will. Indeed, there are inherent limits to any Internal Control system. These limits result from a number of factors, including uncertainties in the outside world, the exercise of judgment, or malfunctions that may occur as a result of human failure or simple error".

These internal control limits can be presented as risks to the company. It is important to be aware of the risks that may arise.

SECTION 3: REVIEWS AND RECOMMENDATIONS

This chapter summarizes the criticisms and recommendations arising from the internal control assessment.

1. CRITICS

a. Internal control weaknesses related to procedures at SOMAGEP-SA :

Weaknesses detected are inherent in the strict application of procedures, with a lack of rigor noted on the part of the company's managers, which increases the risk of failure.

c. Internal control weaknesses linked to the general internal control environment

Weaknesses relating to the agency's general internal control environment were identified:

- Neglect of the value of implementing a procedures manual
- no provision has been made for the remaining unused amounts in the budget

Following a summary of the weaknesses in SOMAGEP-SA's internal control system in order to better understand the appropriate solutions to the problems, recommendations will be formulated to resolve the dysfunctions identified.

2. RECOMMENDATIONS

In view of the malfunctions detected, the recommendations are as follows:

- update the internal control management procedures manual ;
- accounting documents must be checked and signed by the accountants for greater reliability. Documents must not change signatories.
- archives need to be computerized, so it's a good idea to acquire a computer to store archives digitally, so as to be sure of their safe storage.
- Cumulative functions must be taken into account when updating the internal regulations.
- the inspection must be carried out using material (signature, date, name, printout)
- the transmission book needs to be used more regularly.

GENERAL CONCLUSION

It is important to highlight the relevance of the results to our assumptions, however the results of these interpretations reconciled to the pre-established assumptions led to the following results.

Reliable management information can only be achieved through the implementation of internal control procedures capable of accurately capturing all the operations carried out by the organization. The quality of the internal control system is therefore paramount. Internal control is all the more relevant when it is based on rules of conduct and integrity laid down by the governing bodies and communicated to all employees.

Indeed, the internal control system alone cannot prevent SOMAGEP-SA employees from committing fraud, contravening legal or regulatory provisions, or communicating misleading information to the outside world.

In this context, exemplarity is an essential vector for disseminating values within SOMAGEP-SA. Indeed, it cannot be reduced to a purely formal system, in the absence of which serious breaches of business ethics could occur.

Contact with reality provided an opportunity to supplement my theoretical knowledge. Also, the observations made highlighted the strengths and weaknesses of SOMAGEP's management system.

These shortcomings give rise to the problem of weak internal control of the administrative, commercial and accounting management system, and the weakness of the general internal control environment.

A number of solutions were proposed, including: the availability and updating of procedural manuals, the application of these manuals by employees, the reliability and regularity of accounting documents, electronic archive management, the regulation of the separation of duties, and internal regulations should be drawn up to regulate the accumulation of functions and the materialization of the control of sensitive information (signature, date, name, printout).

SOMAGEP needs to start revitalizing its internal control system.

Other avenues which unfortunately could not be explored could be the cause of the problems linked to SOMAGEP's internal control malfunction. Thus, priority must be given to future research, in order to invalidate or confirm the results of the present study, and to explore other useful avenues.

BIBLIOGRAPHY

Book :

Berle A. and Means G. (1932), *The Modern Corporation and Private Property, Transactions Publisher.*

COOPERS, LYBRAND, 2002, La nouvelle pratique du Contrôle Interne. Paris, Edition d'Organisation, 378p.

GRAND B., VERDALLE B., 1999. Audit Comptable et Financier. Paris, Economica, 111p.

GRENIER C., BONNEBOUCHE J, 2003 Auditer et contrôler les activités de l'entreprise. Paris, Edition Foucher, 192p.

PRICEWATERHOUSE, COOPERS, IFACI, 2004. La pratique du contrôle interne. Paris, Edition d'Organisation, 378p.

RENARD J., 2004.Théorie et pratique de l'Audit interne, Edition d'Organisation, 5eme Edition, 486p.

SOMAGEP-SA inventory management procedures PO-014

Purchasing management procedures for PO-015 SOMAGEP-SA

Commercial management procedures PO-005 SOMAGEP-SA invoicing

Commercial management procedures PO-004 SOMAGEP-SA subscription

Commercial management procedures PO-012 SOMAGEP-SA claim

All work instructions attached to the various procedures

SOMAGEP-SA's internal audit charter

Memories

Aksouh Hani and Mehenni Samy Ismail: L'appréciation du controle interne selon le référentiel COSO Licence en sciences commerciales et financières option: Comptabilité 2008. ESG ALGER

Other documents.

Webography

https://fr.wikipedia.org/wiki/Contr%C3%B4le_interne

https://www.manager-go.com/finance/controle-interne.htm

https://www.lecoindesentrepreneurs.fr/controle-interne/

https://amyotgelinas.com/les-avantages-dun-systeme-de-controle-interne-au-sein-de-votre-entreprise/

https://www.ifaci.com/formations/elaborer-le-dispositif-de-controle-interne/

http://www.pansardassocies.com/www/fr/accueil/publications/au dit com ptabilite/points comptables specifiques . aspx.

www.coso.org.

https://www.memoireonline.com/01/09/1913/Analyse-du-controle-interne-au-sein-dune-institution-de-microfinance.html

https://www.etudes-et-analyses.com/theme-economique/controle+internal

https://cgsp.ml/wp-content/uploads/2017/11/7-Evaluation-Controle-interne.pdf

http://documentation.2ie-edu.org/cdi2ie/opac_css/doc_num.php?explnum_id=2636

https://optimiso-group.com/articles/10-etapes-pour-un-controle-interne-efficace/

APPENDICES

QUESTIONNAIRE

1. Are ethical and integrity values formally defined within SOMAGEP?
2. Are the staff carrying out the activities competent?
3. Are the values of ethics and integrity congruent with the company's management styles?
Are responsibilities defined?
5. Do agents have job descriptions?
6. Is there a company organization chart?
7. Is this organization chart regularly updated?
8. Do you have a risk map? If so, how up-to-date is it?
9. Is the company's internal control system assessed?
10. Is internal control a regulatory requirement?
11. Is there an internal procedures manual?
12. Is this manual regularly updated?

13 Does your organization have its own audit charter?
14. Does your organization have an audit department?
15. Does your organization produce an annual internal control report (reporting on internal control measures)?
16. Are these control measures constantly adapted

TABLE OF CONTENTS

Printed by Books on Demand GmbH, Norderstedt / Germany